Unconventional

Marine Carriers

2nd edition

By

Robert H. Vollmerhausen

Table of Contents

Section 1
Integrating components

Section 2
Maritime technology

Section 3
Mechanics and operations

References

Introduction

Unconventional Marine Carriers was first published in 2005. The images were grayscale and the subject of redeveloping maritime transportation was limited in scope. This revised work has new material, full color images and a more comprehensive explanation of the benefits of redeveloping transportation by water.

A theme of this work is to treat maritime transportation as part of a much large more comprehensive transportation concept. That concept is called *'Unconventional Transportation'* that advocates direct 'linking' of one transportation component with another—do away with train stations and airports.

Another theme that is also central to this work, with the addition of new material, is to describe how maritime transportation can and should be instrumental in co-generating much of the electricity needed for equipment operation.

Unconventional Marine Carriers also furthers the thematic concept that the era of 'Happy Motoring' is ending—so unless we are happy walking we should be thinking creatively about transportation and about energy.

Section 1

Integrating components

Part 1

Conventional vs. unconventional transit

1

Oil and conventional travel hubs

Why develop alternative transportation? Isn't oil going to be replaced by biofuels or batteries? Another book *Unconventional Transportation* details essential facts about our energy situation and our overuse of oil with the conclusion that there are no good, plentiful and cheap substitutes for petroleum; and, that without a fuel that equals the energy density of oil, the era of driving, especially between cities, will end. The best way to save energy is not to burn it in the first place.

Less driving means less fuel tax revenue

States are going broke. *Interstate Highways* will most likely not get the trillions of dollars in maintenance required. Potholes and crumbling bridges are in our future.

Declining tax revenues for road and highways means more financial trouble for our aging Interstate system. The collapse of fuel tax revenue for road and bridge repair, along with other economic (and social) problems, will exacerbate our Interstate Highway funding crisis.

Still, assume for the moment that the highways are well-maintained, that the money for upgrades and repair is found; and further, assume that oil continues to flow so that oil demand is satisfied—even with all of that people will still need to fight traffic congestion to get to a train station. They need to find a parking space when they get to the airport.

It's unlikely that all of these conditions: oil flowing, money found for road and bridge maintenance, budget money found for airport and train station maintenance will continue. We have reached the end of the road.

Oil and other fossil fuels are common factors in our existing economic and impending transportation crisis: The cost of oil will prevent many people from driving and the cost of oil will also prevent many people from flying.

It takes hundreds of people to fill the seats of a large modern jet aircraft. And, almost all need to drive to an airport.

What about recession or worse? What if the United States and the world enter into a new era of declining growth and limited resources? If the economy sinks it will be increasingly difficult to find money to maintain huge, conventional transit hubs and highways. Both types of infrastructures, highways and large conventional airports and train stations, will suffer from a lack of adequate funding.

All conventional transportation facilities require that passengers drive to them, because conventional airports, and train stations are *adjuncts* to our Interstates.

In 1850 it made sense to organize transportation in a 'hub' arrangement. Passengers rode in a horse-drawn wagon to a train station, got off the wagon, went into the station, bought a ticket and waited for the train.

Unconventional Marine Carriers

We are now in a world where we cling to our antiquated business methods as if there was something inherently good about train stations, airports and a ticketing process that dates back to the horse and buggy.

Unconventional Transportation suggests that, rather than driving to an airport or a train station, passengers take a bus to start their journey. Each bus has a driver and has a *customer service representative*.

Buses take passengers to a restricted *Passenger Transfer Site*. Passengers transfer from the buses into waiting areas for subsequently boarding of their train, ship or aircraft. (This is an expanded version of the Dulles International 'mobile lounges.')

Buses, storefronts and security

In taking a flight, for example, staff at a local storefront uses biometric measures to identify passengers and pets. Other security measures are used to positively identify all baggage.

The first 'entry' security check is on approaching a van or bus that drives passengers to a local transfer facility. The reason for the 'entry' security check is to insure that the people that passed security in the local storefront are the *same* people getting into the van or bus.

Electronic (biometric) security parameters 'travel' with the passengers. When buses arrive at the passenger transfer facility there is another security check. Methods are used to insure that people have not traded places.

Part 2

Review of Unconventional Transportation

1

Driving between cities will be problematic

Over time petroleum will be pricey and scarce. One alternative to our oil dependency is to develop alternative transit systems that do not operate on fossil fuels.

In *Unconventional Transportation,* I briefly survey the idea that hydrogen, bio-fuels or other technology will keep people *driving* and conclude that, except in very limited situations, the age of unlimited driving will end. If we are going to get from city-to-city it will be by train, air, bus and even water taxis.

This concept is a private—moneymaking—business. Unconventional Transportation (or machine-to-machine transportation) is a concept suggesting using older modes of transportation (trains, buses, trolleys, and even ships) to largely replace driving city-to-city and even suburb-to-city, but to organize those activities in a way that meets the public's transportation needs while making money for investors.

UT also benefits conventional mass transit.

Unconventional Transportation benefits *public* supported mass transit four ways:

1. It does *not* compete with ordinary city/county bus service or metro subways systems for money or patrons.

   There is no 'stealing' patrons nor is there any competition for public money. Public funding that might be available goes to ordinary mass (public) transportation and *not Unconventional Transportation.*

2. UT *does* provide a 'flow' of patrons into public transit systems, because:

3. Unconventional Transportation provides transportation from suburbia to cities. Such transportation services do not 'encroach' on urban public transit facilities. It is supplementary, but not competitive.

4. Commuters from suburbia using a UT system transfer, if they wish, to an urban bus or subway system to complete their commute.

Intercity travel and maritime transportation

Over time the cost and availability of gasoline will make *both* long-distance driving and commercial air travel unrealistic for most people.

Another work *Railroads, Biomass and Synthetic oil* makes a case for growing large areas of biomass along railroad rights-of-way to provide the country with a sustainable source of synthetic oil. This type of program might, if lawyers and public would allow it, to provide a means for supplying aviation fuel to airlines along with a source of diesel fuel for trains; it is unlikely, however, that enough synthetic fuel can be grown to accommodate the public's appetite for gasoline.

Unconventional Marine Carriers

This work concentrates on redeveloping *maritime* transportation for intercity and suburb-to-city commuting. Reasons for including maritime transit in Unconventional Transportation is that:

1. Maritime travel is economical—it provides a 'straight-through' travel route. Linking components means 'connecting' between a bus and a ship is easy—no waiting on platforms.

2. Maritime travel works cooperatively with buses, trolleys, and trains to provide the customer with seamless, integrated travel.

This work suggests a program for saving energy by getting people to voluntarily take buses and trains rather than drive.

Unconventional Transportation uses the expediency of mechanically coupling different transport modalities to create a coherent transportation network.

Linking various transportation components allow passengers to move from one mode of travel to another (bus to train for example) without the use of train stations or waiting platforms—direct machine-to-machine connections.

If *any* technological civilization is to survive the demise of cheap oil it will only be possible with the development of other organizational types of transportation that drastically reduce the use and need for liquid fossil fuels; further, those newer types of transportation must relegate automotive driving to secondary (local) travel.

2

Linking components

> Any transportation system must have a way to safety and conveniently transfer passengers from one means of transport to another.

All conventional transportation also 'links' different types of transportation. Linking is done by way of a 'hub'—such as metro stations, train and bus stations or airport facilities. Public bus systems just use sidewalks or bus shelters where patrons wait for their ride.

Conventionally, the 'connections' passengers make between trains, for example, consists of buildings with waiting platforms: Passengers exit one train, wait on a platform, and then board another train going to their destination.

Brick and mortar buildings are expensive. They are not moveable. If the market or demographics change the metro or subway station stays where it is. Once a metro station is built, used or not, it's in that location forever unless torn down.

Unconventional Transportation safely and conveniently transfers passengers from one transit component to another, but without the use of brick and mortar buildings and without the use of waiting platforms—It directly *links* one transportation component with another. This work illustrates that concept using maritime transportation.

> *Unconventional Marine Carriers* illustrates innovations in maritime travel that are compatible with the concept of linking one component with another so as to give passengers a unified travel experience.

Example 1—Dulles International Airport.

Since the 1960's Dulles International Airport has used 'mobile lounges' to transfer people from terminal buildings out to aircraft waiting on the tarmac. These mobile lounges are large motorized rubber tired vehicles—think of RV's on steroids.

Passengers in the mobile lounges walk from their seats in the mobile lounge into waiting aircraft where flight attendants check their tickets and direct passengers to their seats.

Dulles International Airport's use of mobile lounges to ferry people from air terminals to waiting aircraft is an example of *Unconventional Transportation*.

Benefits include:

    a)   It eliminates much of the need for waiting platforms and other types of auxiliary structures such as covered walkways—passengers ride rather than wait and walk to their aircraft.

    b)   It's economical and safe. Passengers in the mobile lounge are protected from weather, from heat and cold, and transfer directly from the mobile lounge into the aircraft.

    c)   It's flexible. Linking transportation components such as air terminals and aircraft with mobile lounges provides a flexible means to safely and cost-effectively transfer passengers to and from aircraft.

Example 2—trains

Somebody realized a hundred and fifty years ago that it was much more efficient to *link* railcars together than to drive each railcar, as a separate conveyance, down the rails.

Railcars are linked, or coupled, so that passengers move from one railcar to another by going out one door, crossing over the connector, and walking into the next car.

Unconventional Transportation is another way of *linking* transit components. Fundamentally, people should transfer *directly between different transit components* (machine-to-machine) rather than wait on sidewalks or in bus shelters.

This work concentrates on maritime applications, but each device or machine is compatible with the overall concept of linking transit components. In this work 'linking' is implied or inherent in the design of the machines.

Another goal of this work, along with describing how a linking methodology can function efficiently is to present two new innovations in maritime high-speed travel:

1. A Hover-effect ship for high-speed maritime travel over protected waterways, and

2. Automated marine shuttles (AMS) that operationally integrate into Passenger Transfer Sites for moving passengers quickly, safely and comfortably from one type of transportation (a water taxi) to another mode of travel such as a bus or a train.

3

Integrated components. One system

Maritime transportation is an economical addition to a total transit system, because it fits the profile of transportation elements that:

1. Work cooperatively together, and

2. As with buses and trolleys maritime travel requires little in the way of infrastructure—ships travel over protected waterways. Only small landscaped *Passenger Transfer Sites* are required: No public parking lots, no ticket counters, no buildings, except for passenger accommodations.

> Passengers arrive on a bus and transfer directly to a water taxis or other maritime means of transportation. They are transported across water and transfer directly from a water taxis (or ship) to *another set of buses* to continue their commute.

All components lend themselves to economical operations, because heavy construction isn't needed such as overhead electrical wires, or other expensive infrastructure. No tunnels. No massive construction.

The buses, water taxis, and trolleys that make up the transit concept function together cooperatively, but do not require extensive auxiliary construction such as terminal buildings, ticket counters or parking lots.

Maritime transportation is economical even as a 'stand-alone' system, but much more economical as part of a total transit network.

4

Geographical size and its implications for alternative transportation and energy

Another reason for redeveloping maritime transportation is the large size of the United States. The country's size, given its financial situation, *limits its options* in the types of transportation that can be built and maintained.

The United States has very limited policy choices, because those choices are limited to developing transit and energy infrastructures that *do not* require billions of dollars in new expenditures; that is, any new infrastructure must be *cheap—or undertaken by private money*.

Maritime transportation is economical in that there are no large extensive construction expenditures, such as terminal buildings, bridges or tunnels to be built. Maritime transit is also economical in its energy requirements in that almost any fuel can be used; and, unconventional transportation provides a 'built-in' energy distribution infrastructure. That aspect is discussed in Part 3 of Section 3 starting on page 62.

The United States is large and its financial resources are depressingly small. The debtor status of the United States limits the kinds of transportation that can be developed.

This work introduces maritime innovations for redeveloping travel by ship and by water taxi. Although not expressly illustrated all of these innovations are part of the larger, component integrated transportation concept.

5

Maritime transit as part of a total transit system

Maritime transportation, on protected water, provides a potential 'straight-across' routing to save time and energy in the transportation process.

Redeveloping maritime travel means that, rather than follow the roads and highways around to a bridge (that might be in dire need of repair) passengers merely transfer from buses to water-taxis and continue on their way—straight across the river to another set of buses. This is one benefit of 'linking'—straight-through travel.

People directly transfer from buses to water taxis—and then back to buses—is the idea of linking transportation components to create total transit. Americans, however, love speed so the 'water taxis' must be modern, safe, comfortable and *fast*.

Most of the innovations introduced are a combination of watercraft and ground-effect machines. They 'skirt' the surface of the water much like a low-flying aircraft and stay above the surface by compressing air downward through vents.

One of the difficulties in designing these machines is in achieving easy control and responsive flight characteristics along with flight stability. Part 1 in section 2 provides brief history and background of hovercraft and low-flying machines.

Section 2

Maritime technology

Part 1

Flight stability—a brief history

1

Background and history

As far back as World War One airplane designers understood that flying qualities are a difficult blend of competing forces. Some aircraft were built and flown that were so stable in straight-and-level flight that the pilot could not make the airplane turn. The airplane only flew straight.

Other aircraft were built that the pilot couldn't take his hand off the stick. The airplane wobbled around and fell out of the air. These aircraft were so unstable that the pilot had to fight to keep it flying straight and level. All of the stability problems associated with new aircraft design are also true for ground-effect machines.

Traveling at high speed over water is inherently dangerous, because the pilot does not have time to react if something's wrong. Pilots have a saying that *there is safety above and danger below.* While traveling close to water's surface human reaction time is too slow to effectively correct an impending accident. By the time a pilot reacts to an event, realizing that something is wrong, the machine has crashed.

Stability, control and maneuverability depend to a large extent on a machine's inherent design qualities. Inherent stability has to do with the ability of the machine, without any corrective response from the pilot, to return to an equilibrium flight condition after a transitory disturbance.

Wing-in-Ground Machines

Safety above and danger below is especially true of Wing-in-Ground (WIG) machines that utilize aircraft type wings to generate much of the aerodynamic lift required to keep the machine airborne. If a wingtip enters the water, the WIG machine will probably cartwheel into a crash.

Hovercraft

A hovercraft is also called an air-cushion vehicle, because air is compressed under the apparatus that forces the machine off of the surface. A hovercraft floats or rides on compressed air as *aircraft* engines drive the machine forward. Conventional hovercraft are not used much around cities, because of their engine noise.

Hydrofoil ships

Hydrofoil ships use hydro-skis to generate a coefficient of hydrodynamic lift. They achieve high speeds, but the ride is often rough. Hydrofoil ships are also noisy and expensive to operate.

Hydrofoil ships of 'conventional' design suffer from the same cost/ performance defects as the supersonic Concorde aircraft—they don't carry enough passengers to be consistently profitable.

Hover-effect ships

Based on the author's U.S. patent # 6,497,189 issued December 24, 2002, the Hover-effect ship is a combination between a conventional hovercraft and a hydrofoil ship.

When looking at illustrations of the proposed *Hover-effect Ship* is 'Why so big?' The short answer is that it's necessary. As automotive driving drifts into history, as people cannot afford gasoline or the States can't afford to upgrade and repair aging Interstate highways, other cheaper transportation must be found. And, these other transportation resources must provide the carrying capability to move many people economically, conveniently and safety.

The United States has plentiful water resources. The potential for reestablishing large-scale maritime transit, however, must take place within the context of consistent profitability; and, provide enough seats at a price that people can afford.

Year around operation—on protected water.

One major factor in providing profitable maritime operation is to have *year-around* service. The equipment illustrated in this work has operational capability for all-season use, *although how these components function may differ from season to season;* that is, watercraft may be displacement hulled watercraft one season and operate as hovercraft—skimming over the surface—in winter.

Part 2 describes the Hover-effect ship in general terms. A more complete mechanical description is provided in part 3.

Part 2

Hover-effect Ships

1

Large Hover-effect ships

A Hover-effect ship (or smaller Hover-effect Craft) is a patented innovation. The design and operational characteristics this innovation is described in this and subsequent sections.

The illustration on page 17 shows a view of a Hover-effect ship with front-mounted propulsion pods that *pulls* the ship through the water. The front-mounted hydrofoils are part of a pair of propulsion pods that provide hydrodynamic lift while moving the ship forward.

Hover-effect ships use a two in front hydrofoil arrangement—two hydrofoil elements in the front and then one (substantially centered) in the stern so as to provide the ship with 'three point suspension' when traveling too slowly to utilize hydrodynamic and aerodynamic lift.

The rear hydrofoil is retractable and used when the ship is traveling slowly or maneuvering. At cruising speed the stern hydrofoil retracts to improve laminar airflow under the ship.

Using modern materials, construction techniques and aircraft power plants Hover-effect ships can be built to carry from three hundred to five hundred people at speeds up to sixty to seventy knots.

Hover-effect ship passenger cabin

A passenger cabin structure rides on top of a wide platform assembly. The passenger cabin (superstructure) 'floats' on a *pneumatic suspension system*. The passenger cabin is 'isolated' from the motion of the undercarriage. The passenger compartment is also *mechanically* separated from the front driving engines by hydraulic suspension modules.

Inflatable pneumatic elements located in the center of the platform assembly function to support the passenger cabin—cushion the ride. The suspension system, pneumatic and hydraulic, dampens vibrations set up by the passage of the ship through water.

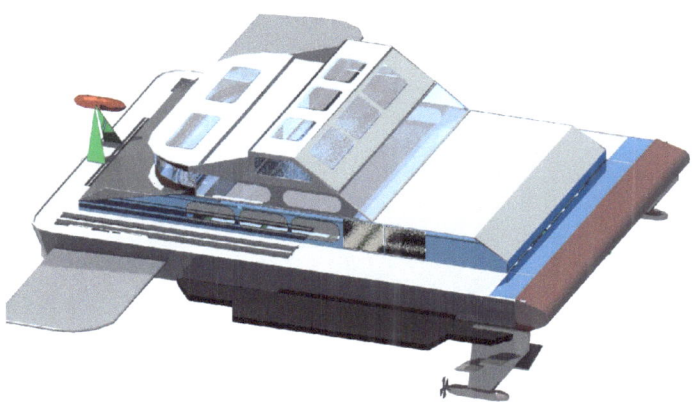

The passenger cabin structure is straddled between the bow (hydrodynamic lift on the engine pods) and the stern that generates a coefficient of aerodynamic lift. Bow and stern have *different* lifting mediums—water in front and air at the stern.

Both *aerodynamic and hydrodynamic* lift is generated to 'cradle' the center of gravity at a point amidships so that the ship is both smooth riding and also maintains a stable flight configuration.

The passenger compartment rides on a *platform assembly* illustrated below. It shows a center-mounted *pneumatic island structure* with an array of individually controllable pneumatic cushion elements that support the passenger compartment. The cushion elements are shown in brown in the illustration below.

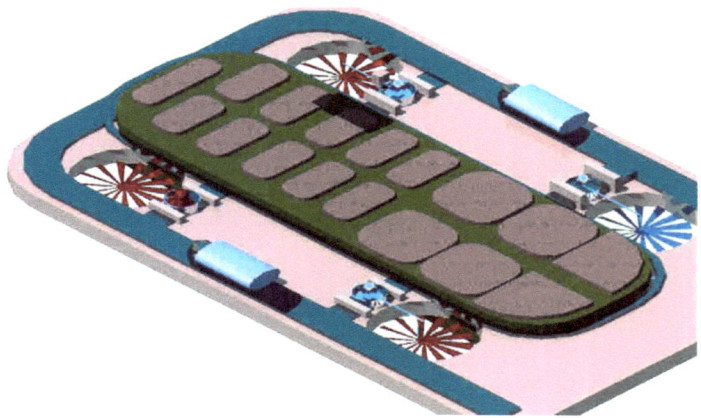

Pneumatic elements support and position the passenger compartment; and, by selectively inflating these pneumatic elements (centered in the illustration above) the passenger compartment is *elevated* along the longitudinal and the roll axis as, for example, in a turn.

In this way the passenger compartment is 'tipped' into a *coordinated turn* for passenger comfort. Pneumatic elements inflate so as to lift one side of the passenger compartment in turning and maneuvering.

Along the platform periphery circular cutouts support a series of engines to compress air under the ship in a *hovercraft* operational mode.

Air is compressed under the ship. Energy of the ground effect is used to buoy the machine upward off of the water's surface as the ship reaches cruising speed.

Mechanical stability

The design of the wing assembly at the stern controls the motion of the ship in the *pitch axis*. If the angle of attack of the wings changes, as by the machine going down by the bow, the lift generated on the wing is decreased. Less lift at the back causes the stern of the Hover-effect machine to lower, temporarily leveling the machine at a lower dynamic flight level.

The Hover-effect ship is stable in the *roll axis* due to the machine's wide stance. Waves pass underneath the machine while at cruising speed. Furthermore, *air is compressed* under the Hover-effect ship as it travels forward.

The design creates laminar airflows under the machine so as to counteract wind pressing against the Hover-effect ship's walls. It flies on a cushion of air.

Comfortable ride

Hover-effect Ship passengers are separated from the water by:

❖ The compressive effect of the ground effect under the machine;

❖ Hydraulic suspension within the (front) engine cowlings;

❖ Pneumatic suspension elements separating the superstructure from the platform assembly.

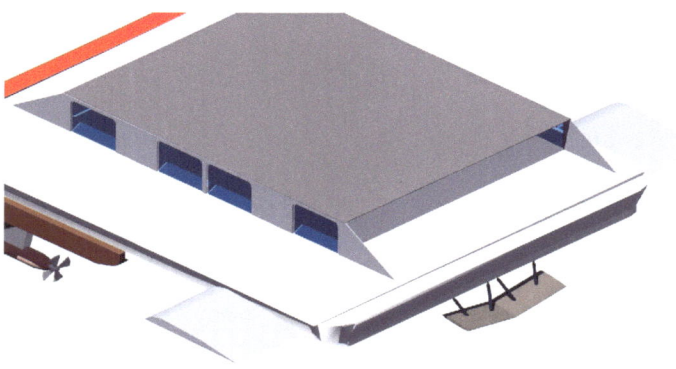

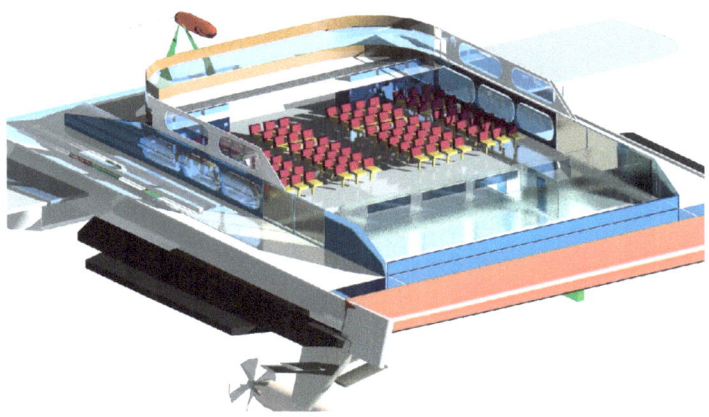

A smooth ride even at high speeds of seventy miles knots makes other amenities possible such as restaurants or even a movie theater possible. These amenities are possible, because Hover-effect ships would also be *quiet* and smooth riding—the driving engines are in the water.

Passengers hear the movie and not constant noise from the engines. Passengers don't hear roar of propellers, because the props or jet thrusters are *in the water and* not pushing air as in conventional hovercraft.

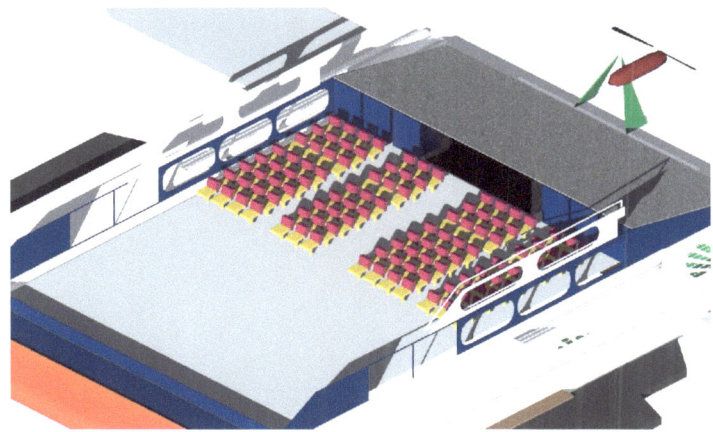

The Hover-effect ship's wide 'stance' creates a platform for enclosed observation decks, restaurants, childcare, and other amenities found on large cruise ships.

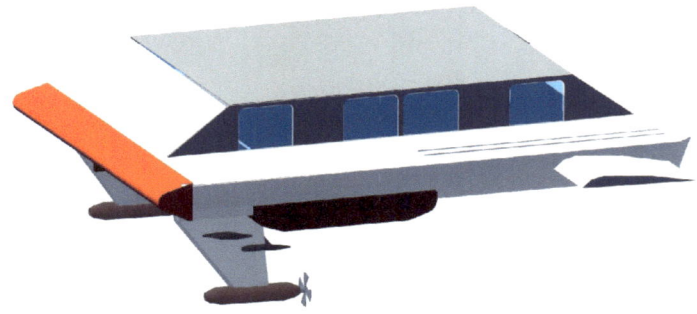

Economy is essential so the unit cost per seat must be kept low. The reason for this is that the likelihood of government financial subsidy is substantially zero—even if various branches of government are not broke. Economy must be matched, however, with quality service and quality service includes a comfortable ride.

The Hover-effect ship offers quality service at an economical price. While large Hover-effect Ships could accommodate as many people as a jumbo jet they would provide transportation service directly to and from major city harbors.

Smaller Hover-effect machines are medium distance water taxis. The next section briefly describes how smaller Hover-effect machines can be used as high-speed maritime taxis.

2

Hover-effect *craft* as water taxis

The illustration shows a smaller Hover-effect machine for transporting thirty to forty passengers over fairly short distances of twenty miles or so. The wide platform assembly contains engines for venting compressed air below the machine for operation in its hovercraft mode.

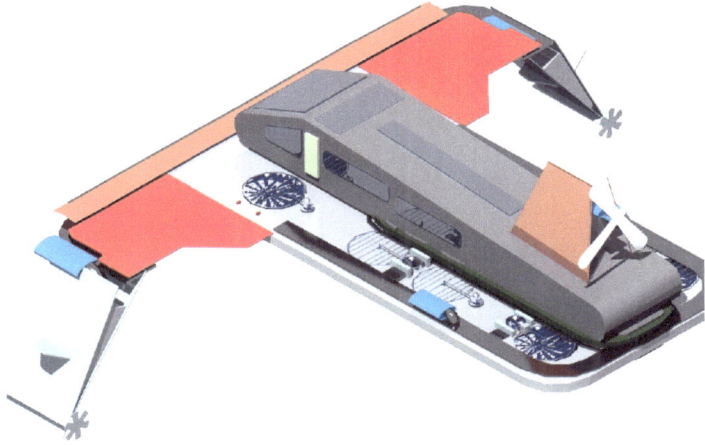

These machines are fast and maneuverable and have straight sides for convenient transfer of passengers to other transportation modalities such as cable car trolleys. The illustration on page 24 shows two cable car trolleys suspended on either side of a water taxi so that passengers may *exit* to one trolley while *others enter the water taxi* from the other side.

All of these maritime machines are part of a total transportation system that uses trolleys and buses as key transit components.

Maritime transportation can be developed that moves people quickly and easily up and down bays and across rivers to connect with buses and other transportation.

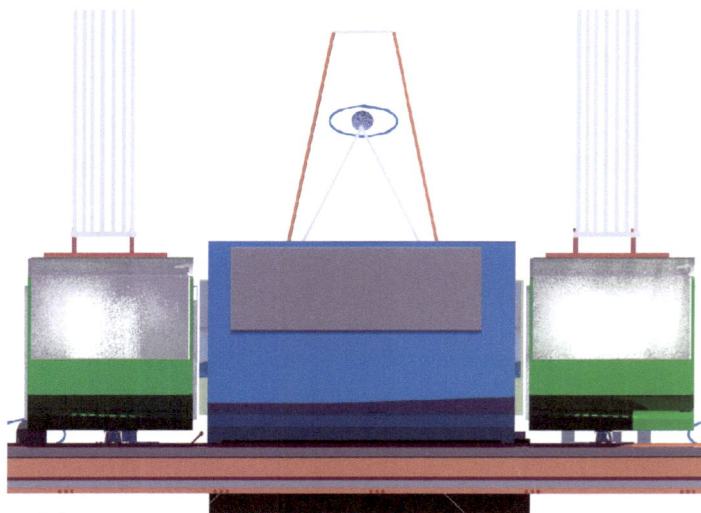

This concept is to use both high-speed ships and connector technology, such as the cable trolleys, to integrate interurban travel into a convenient, economical and safe transportation schema. High-speed water taxis are one component of integrated, component linking, transportation.

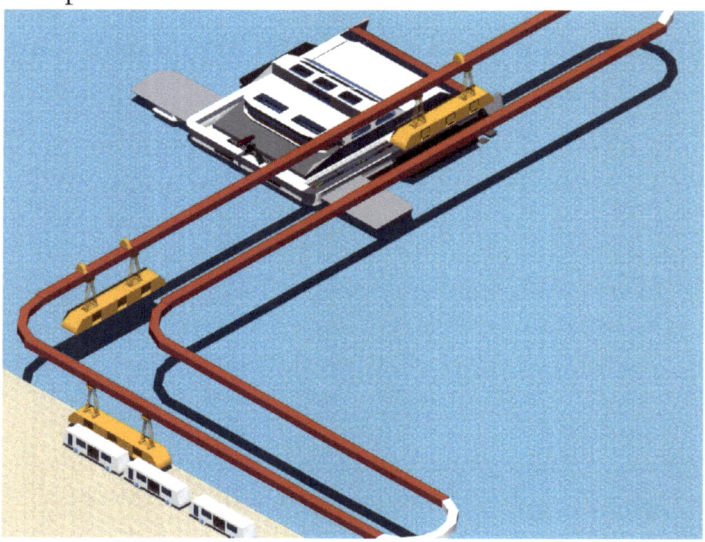

Part 3

Mechanics of Hover-effect ships

1

Multiple suspension systems smooth the ride

Just as there are multiple suspension means designed into the Hover-effect ship to cushion the ride there are also multiple means for generating lift.

Multiple suspension systems and multiple lift systems *function cooperatively* to smooth and mechanically stabilize the ride.

Hover-effect ships, like all boats and ships, are subject to disequilibria by both wind and wave. All craft are designed to function in the aquatic and atmospheric environment.

Riding characteristics of a Hover-effect Ship is a composite of the different types of lifting forces that lift the machine. Waves and wind gusts, water and air, affect the ride differently.

By generating a coefficient of *aerodynamic lift* at the stern, and *hydrodynamic lift* at the front Hover-effect ships have a dynamic of motion determined by *completely different sets of lifting forces*, in addition to *ground-effect* lift under the wide platform assembly.

Hydro-skis, attached to the front engine assemblies, flare outward like wings. The effect of this design is to increase the lift of the skis if the nose dips downward in the pitch axis.

Unlike a conventional aircraft, the stern wing assembly does *not* assist the Hover-effect Ship in making coordinated turning maneuvers, but only provides aerodynamic lift, to support the ship's weight and provide increased flight stability at cruising speed.

At cruising speed a Hover-effect ship is 'cradled' with lift generated bow and stern, so that the Center of Gravity would have to displace forward, over the front engines, for a Hover-ship to 'nosedive' into the water. That event would be improbable.

The placement of the Center of Gravity makes Hover-effect ships very stable, and yet responsive to the pilot's actions while, at the same time, not giving the pilot enough control to drive the machine down into the water.

Flight stability, maneuverability, and safety are intertwined, and find expression in the right design. The motion of the ship is a complex blend of these different lifting forces that, at the right speed and heading would tend to cancel out periodic vibrations and cyclic motion of the ship resulting in a smoother and more mechanically stable ride.

The design, using different means for generating lift, creates a self-regulating dynamic resulting in a stable flight configuration. All of the design elements coordinate to contribute to a dynamic balance and stable flight configuration of the ship.

Hover and maneuvering

Wave action or currents apply very little force against most of the ship's superstructure. Wind acting against the sides of the Hover-effect ship is counter-balanced by actuation of the stern aero-engine; or, actuation of one or more air compression engines arrayed on the platform assembly.

The benefit of this is that most passengers want to face the direction of travel. Hover-effect Ships would not have to 'crab into' the wind to maintain a specific directional heading. Its undercarriage engines counteracts cross winds to keep the ship pointed in the direction of travel. This uses more fuel but makes for a more pleasant ride.

Hover-effect ships operate by compressing air under the wide platform assembly so as to use the ground effect to lift the ship while it is operating at low speed.

Flight stability at slow speed or standing (hover)

Standing, or at low speed, Hover-effect ships can hover, much as a helicopter or a hovercraft by generating high-pressure air below the platform assembly.

At slow speed engines arranged around the undercarriage provide a downdraft of high-pressure air to generate an air cushion under the Hover-effect Craft.

At hover, the platform assembly lifts off the water with the help auxiliary engines. This is the hovercraft mode of operation.

The Hover-effect Ship has a plurality of inflatable elements arrayed along the undercarriage so that the vessel floats when stopped. These inflatable elements are retractable (much like landing gear on aircraft) so that as the ship increases speed the inflatable elements are deflated and redeployed into the undercarriage as the ship gains speed.

A Hover-effect ship has two modes of operation:

(1)   At slow speed, the Hover-effect ships hovers by compressing air below the platform assembly and adjusting the buoyancy of the engine cowling assemblies. In this regard Hover-effect Ships are similar, although not identical, to hovercraft.

(2)   At cruise speed air is directed under the platform assembly and the stern wing assembly supplements lift.

Hover lift-Slow speed operation

This type of machine has potential for maneuverability and great control as well as stability under high-speed flight conditions.

As a Hover-effect Ship slows down it looses aerodynamic lift. The stern wing assembly doesn't produce enough lift at slow speeds to support the platform assembly and compression-fan engines are engaged to supply lift as the HEM slows down. As the vessel slows the pneumatic floatation gear comes in contact with the water.

As the ship slows down engine cowling assemblies have ballast tanks that flood so that the Hover-effect ship flies down to the water in a level flight configuration.

Linking to high-speed rail is one possibility for effectively using Hover-effect technology. This transportation concept would enable potential customers that are geographically scattered over many counties of Virginia and Eastern Maryland, for example, to take advantage of other transportation such as high-speed rail service.

One advantage of this type of transportation is that it can be used in combination with other maritime and 'unconventional' transit systems to create a very efficient, non-fossil fueled, transportation system.

Component compatibility

One of the major contributions of the Industrial Revolution was that of interchangeable parts. This concept merely takes the idea of interchangeable parts and applies that idea systematically.

Part 4

Hover-effect carriers

1

Carrying shuttles piggyback

The illustration shows the design possibilities of 'linking' machines—It's an alternative to building more parking lots and massive brick and mortar buildings.

One machine carrying another is nothing new. Cargo aircraft carry tanks and buses. Trains carry automobiles. (A747 jet carries the Space Shuttle between California and Florida.)

A modified Hover-effect ship is used as a maritime *carrier*. A water taxi is departing (launching from) the Hover-effect ship.

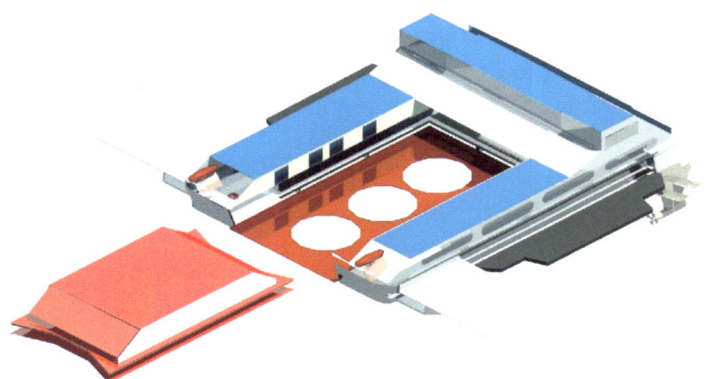

The shuttle locks into the Hover-effect ship docking bay when the two machines travel together; however, the water taxi 'supports itself' by 'pulling air' through its bow and stern vents and directing volumetric air downward—it carries much of its own weight.

In this concept the weight of the shuttle riding piggyback is considered to be too heavy for the hover-effect ship to simply carry. The two machines fly together with *both machines* providing lift.

Passengers transfer to and from the shuttle, but only as the Hover-effect ship is *stopped* or otherwise secured.

This 'carry' arrangement can be modified to work with a variety of different shuttles or other types of equipment. The benefit of this arrangement is that passengers transfer to and from the Hover-effect ship— travel to different destinations without transferring at brick and mortar transit terminal buildings.

Configuration changes

The profile view (above) of a Hover-effect ship transporting a shuttle illustrates how the shuttle 'sits' forward so as to maintain the Hover-effect ship's center of gravity. The two machines are designed to *fly together.*

2

Versatile carrier platforms

Here is another carrier design with the carrier open in the front. This design provides a large bay area for berthing a shuttle for long distance and high-speed maritime travel.

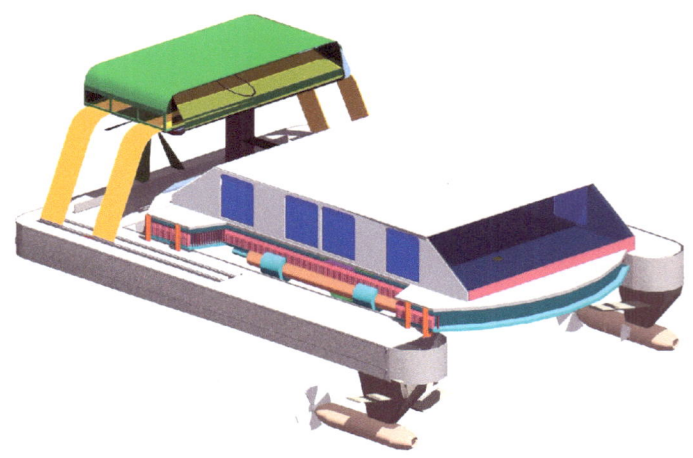

The bridge is at the carrier's stern and high so as to allow the crew complete visibility over the shuttle. As discussed previously the Hover-ship has multiple means of lift so that the ride is smooth.

Operations and control

Water taxi shuttles have *omni-directional water jets* to propel and steer. Shuttle undercarriage also has air thruster engines to generate a *hovercraft air cushion* when partnered with a Hover-effect ship.

The front-mounted hydrofoils are the Hover-effect ship's only direct contact with the water. The hydrofoil housing is hydraulically suspended to buffer and smooth mechanical vibration.

A plurality of air-mover engines disposed at the periphery of the platform assembly provides 'hovercraft' type lift to balance the hydrodynamic lift generated by the front-mounted hydroplanes mounted on the propulsion pods.

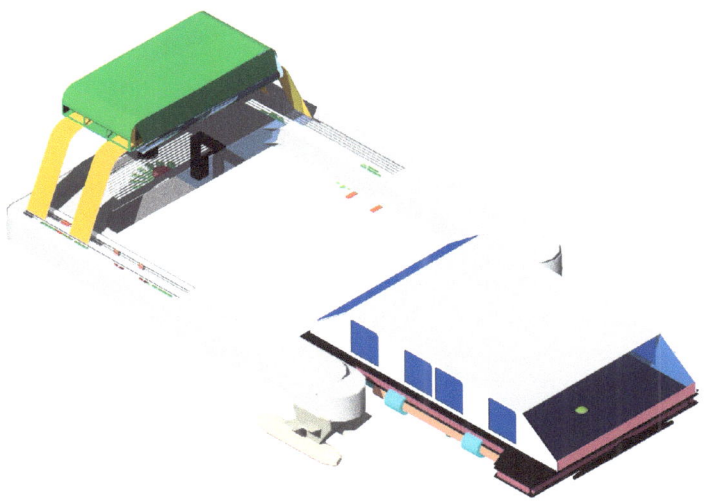

A plurality of engines generates lift and forward thrust. The transport carrier provides a versatile and maneuverable machine for high-speed transit service on protected waterways such as the Chesapeake Bay.

Part 5

Hover-effect utility carriers

1

General description

There are many possible shapes, sizes, and designs of Hover-effect devices. This part illustrates a design for a 'switch engine' application such as ferrying equipment or small amphibious aircraft from one part of an operational (a maritime transportation support area) to another.

There are two forward-mounted hydrofoil assemblies with dual engine assemblies attached so as to draw—pull—the machine through the water. This is a standard features on Hover-effect ships.

The illustration shows a Hover-effect utility carrier that uses its *two elongated wings* as a support base for carrying cargo piggyback.

The operational principle of this machine is the same as for other Hover-effect machines. Two or more *hydrofoil components* are 'front-mounted' and provide motive power by pushing water rather than air.

Each wing also has an engine, substantially as shown, to vent ('jet') air below the wing so as to *elevate and level the wing* as the machine slows down.

As speed increases *aerodynamic lift* provides part of the lifting force to elevate (level) the wings while the stern-mounted engines provide additional vertical thrust.

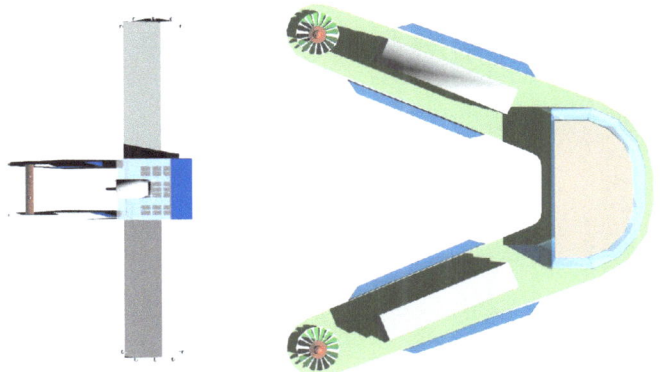

This top view shows an aircraft approaching the carrier from the stern. With the stern engines off the back of the wings are on the surface (drag behind) so as to provide a slanted ramp for the aircraft.

The load/unload procedure 'contains' the cargo (in this case an amphibious aircraft) until it is free of the carrier and able to proceed on its own power. The aircraft is 'corralled' in the slanting wing configuration until the aircraft is under its own power.

Loading and unloading of cargo would be done in a protected environment for providing a level water surface free of high wind interference for carrier/payload operations.

The illustration shows an amphibious aircraft riding on the carrier by hydraulic jacks that hold the aircraft in place. An aircraft moves into the 'bay' and is positioned by a plurality of flexible support elements (shown as highlighted blocks on the carrier's wings).

Carriers include multiple lifting and suspension systems that provide a stable riding platform even in windy conditions. The ride in the elevated bridge structure would be comfortable under most operating conditions.

As with other Hover-effect designs the Hover-effect Utility Carrier can be quite large. In the model shown below the bridge is quite roomy and may, for example, have a restaurant or observation decks or possible individual staterooms for passengers on longer commutes on protected waterways.

The forward mounted engines pull the utility carrier through the water while the wings 'kite' behind until the wing engines generate enough thrust lift to push the wings clear of the water.

Placing the *wing engines* far back on the wing gives the engines 'mechanical advantage' or leverage. The result is that most of the mechanical energy that is required to level the wings is supplied through the in-water front mounted engines that generates an airflow under the machine. Engines on the wingtips only level the wings when the carrier is moving too slowly for aerodynamic lift (ground-effect) to be effective.

Section 3

Mechanics and operations

Part 1

Automated maritime shuttles

1

Water taxi fleet operations

Water taxis that carry two hundred to three hundred passengers across rivers or protected waterways are another maritime appliance that is compatible with the idea of 'machine-to-machine' transportation.

Water taxis—Automated Marine Shuttles—follow a line-of-sight course from one **dock structure** to another. Each shuttle would have its own lane of travel and the 'crossing' would be very quick, perhaps five minutes or less.

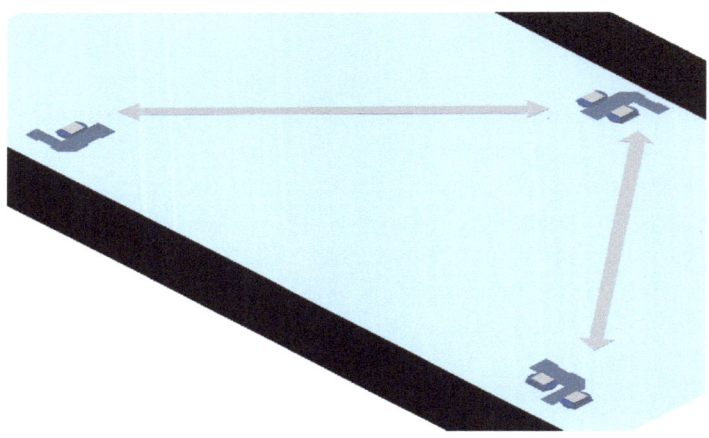

Shuttles have a very simple course to follow, because they are *robotic appliances*. They travel back and forth in a straight line between one *Passenger Transfer Site* (dock structure*)* and another that would be located (usually) diagonally across a river.

AMS takes commuters across rivers, but not necessarily straight across; that is, the water taxis would operate in straight lines, but travel diagonally. Routes would be determined by 'weighing' areas, with the greatest weight given to areas that are *destinations* for the most people.

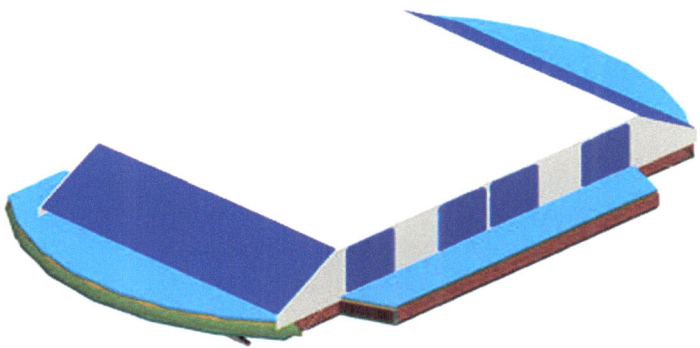

Safety-AMS function as a mutual safety net

Design of Robotic (automated) Maritime Shuttlecraft have two objectives: First, safety of passengers; and secondly, a high level of reliability. Ships are not like buses. If a bus has a mechanical breakdown it rolls to a stop. A ship doesn't stop; it drifts with the current.

Each time a taxi docks each *major operating sub-system* is checked automatically. Fuel, engine temperature and so forth are routinely checked to insure that the shuttle's mechanical systems are operating correctly. This safety check is done through an automated procedure.

Equipment operation changes to meet conditions

A basic design concept is to 'change out' the equipment season-by-season to always provide the most efficient and safest transportation.

The illustration shows a maritime shuttle undercarriage platform with engines mounted in the structure so as to provide a fraction of 'hovercraft' lift in heavy weather.

An 'air' undercarriage, with an array of engines, vents air underneath the machine so as to elevate the machine higher in the water. This is referred to as the 'hovercraft' mode.

In high wave conditions the additional lift provides a 'hovercraft' style ride. Thus the shuttle has two different operational *watercraft* modes depending on season and operating conditions.

When the 'omni jet' undercarriage is *removed* the water taxi is used as a *hovercraft*. This twin propulsion machine is very flexible and can find wide use in many maritime environments. This is shown in the illustration on page 41.

Underneath the 'air' undercarriage is a 'water' undercarriage that support multiple water-thrusting engines (omni-directional jets) for primary propulsion during the boating season. The jets are *blue* in the graphic below.

The air cushion apparatus rides 'piggy-back' on the water jet undercarriage. The illustration also shows the 'hovercraft' operational components of the shuttle separated out from the water jet undercarriage.

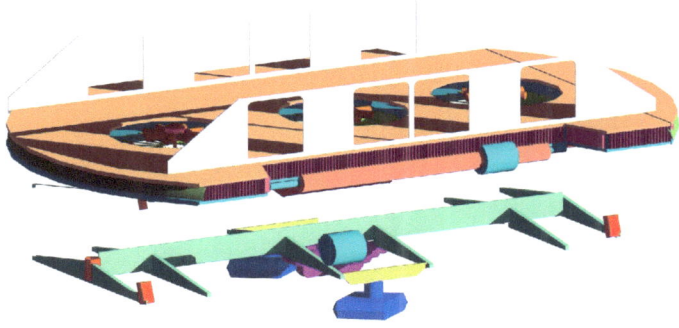

Shuttle—hovercraft operation differs from conventional hovercraft in that volumetric air is ducted into the machine from the shuttle's sides. The shuttles are two-ships-in-one with the capability of switching from one mode of operation to another: displacement hull, semi-hovercraft, or 'straight' hovercraft, depending on weather and water conditions.

Water taxis are part of a much larger, more inclusive maritime transit system as illustrated in the following pages. Maritime transit is part of a total transportation concept in which people move machine-to-machine.

Maritime transit components

Water taxis (Automated Maritime Shuttles) are designed and built to *interlock* into a *geometric docking structure*.

The water taxis 'fit' into the dock in only one way so that the water taxis are always *oriented correctly* so that trolleys operating on a rigid guideway can position (side by side) with a water taxi superstructure for the subsequent transfer of passengers.

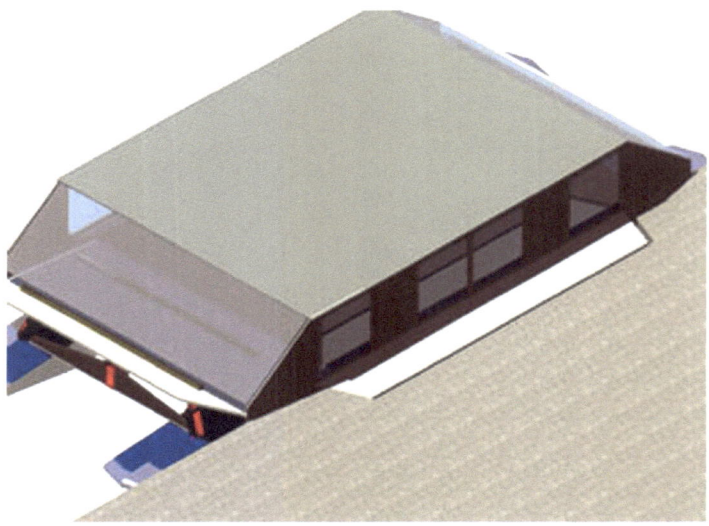

The end result is that all of the components fit together—in all dimensions—for the safe, convenient, and easy transfers of passengers to and from the water taxis.

Engineering of maritime transportation is directed to provide all of the amenities and operational functionality of conventional 'brick and mortar' transit infrastructures, but do so much more cost-efficiently.

The graphic on the previous page shows a Robotic AMS connected to a dock structure by mechanically fitting the shuttle into an interlocking slot in the dock. The slot design automatically aligns the watercraft to and positions it with the dock structure.

Orienting the shuttle while the wind is blowing at night is the primary reason why the docking, and the overall operation of the shuttle fleet, is an automated procedure—expecting human pilots to do this is to expect too much.

Water taxis align automatically with a cable car, because the equipment, *the whole installation,* is designed with components that fit together.

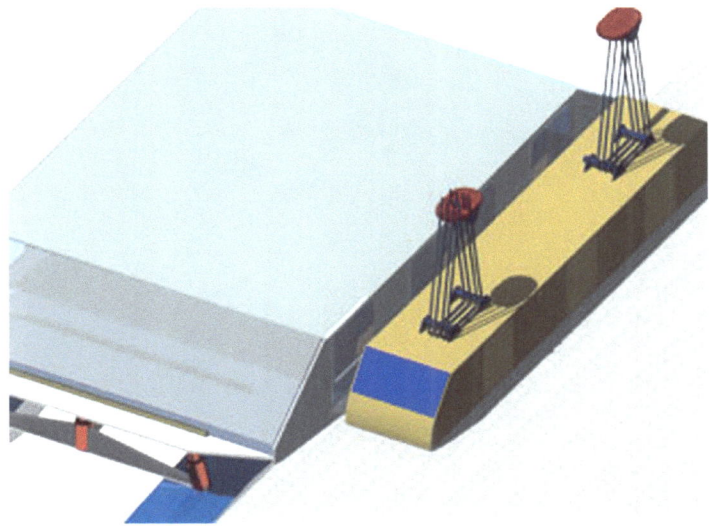

Docks are fairly simple geometric structures that hydraulically rise and fall with river water elevation so as to maintain a freeboard clearance with the water. The cable trolleys are also under program control so as to 'mate' correctly with watercraft in the dock.

2

Bus to water taxi transportation—more linking

A *Passenger Transfer Site*, such as illustrated below, secures Automated Maritime Shuttles by docking the shuttles. The dock structures are engineered to 'rise and fall' with water elevation, but are otherwise stationary so that cable car trolleys align with the sides of the water taxis at the dock.

Passengers transfer from buses to a guideway trolley and then ride out on overhead rails to waiting water taxis. This concept is illustrated below.

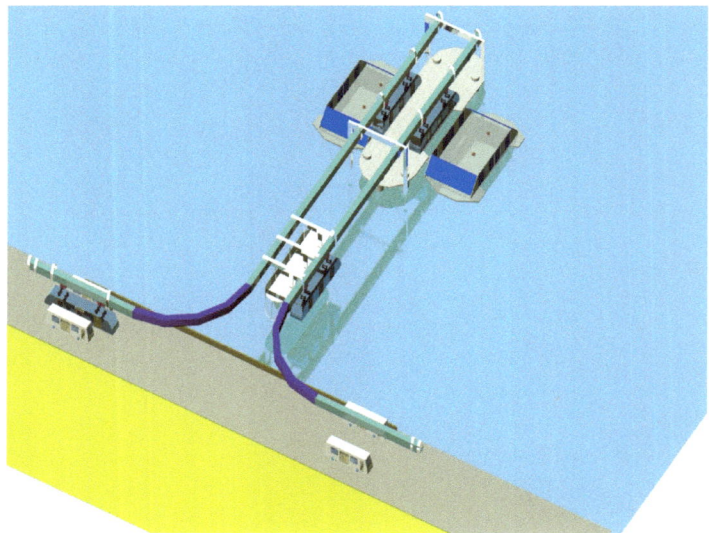

Buses bring passengers to a 'Passenger Transfer Site' where the passengers transfer from the buses to an overhead cable car trolley. The trolley transfers passengers out to waiting water taxis that are docked at the special dock structures.

This concept allows passengers to take a variety of subsequent transportation components that meets their own transportation needs—passengers can transfer, for example, from the water taxis to buses on the other side of the river or bay—or to another ship.

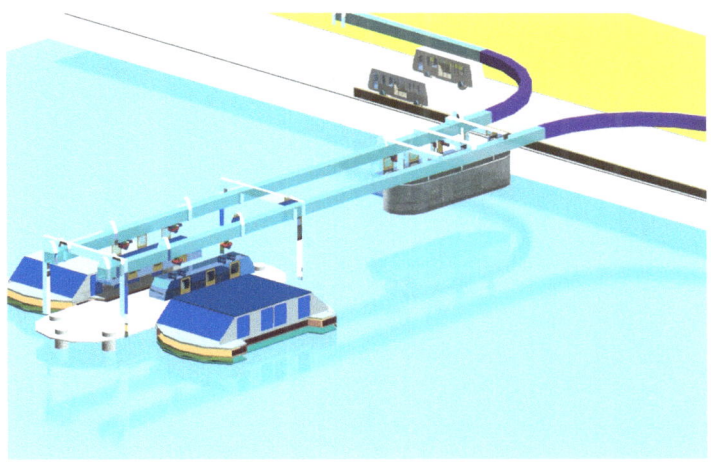

In this concept passengers move 'seamlessly' from one mode of travel to another—from buses to monorail trolleys and then to water taxis.

The object suspended between the guideway near the shore represents a secure *waiting area for passengers*. It may include a restaurant.

Maritime transit and Hover-effect Ships

Dock structures accommodate many types of machines: Hover-effect ships, and water taxis. Machine-to-machine transportation supports a wide variety of transit applications—all with direct connection capability.

3

Direct component linking

Direct connection is a special form of linking. Each component 'pushes' directly against another so as to create an impromptu interconnection passageway.

The illustration below shows two water taxis that have two-passenger compartment laterally moved (shifted) so as to use shuttle-docking geometry to connect one taxi with one another.

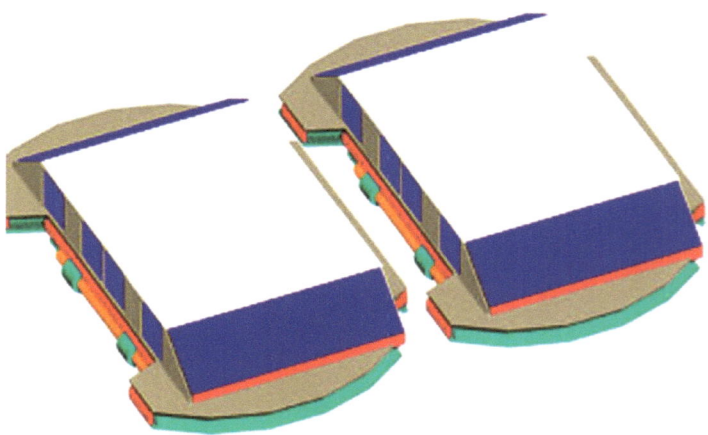

In this way two, or more, taxis, can be coupled—using platform geometry so as to allow safe and convenient passenger transfer between two machines.

This concept of direct component coupling can be extended as shown in the next illustrations to include a 'mother' ship that acts as a *mobile docking structure* while water taxis come and go.

The shuttles in the illustration on this page are docked to a *transport ship* that follows a route up and down a stretch of river. This type of 'mobile dock' allows passengers to move freely between the various docked water taxis—all moving together.

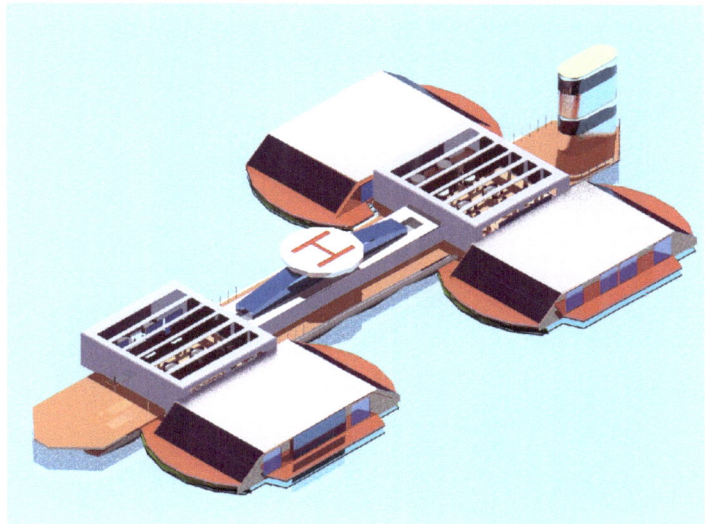

The 'dock-ship' has multiple docking spaces and a long linear passageway to allow passengers to move the length of the ship.

In both the bow and stern, in this example, the dock-ship has large superstructures. These are waiting areas, lounges. There may be a wet bar or a restaurant. The ship may include shops and other amenities.

The ship might also have other amenities such as observation decks, or in the case of an urban emergency, may have *medical facilities*. The entire fleet could be equipped to double as an urban medical Corp —or to evacuate people from danger.

In the event of urban emergency traffic lights on surface roads may be inoperable and the surface roads jammed with traffic.

The United States has abundant water resources. These resources could be used for routine transportation, but also used in emergencies.

Each water taxi may be used as a water ambulance in the event of urban disaster or terrorist attack while the ship used as a mobile hospital. Water taxis can be built with reinforced roofs so that helicopter air ambulances land directly on the roof.

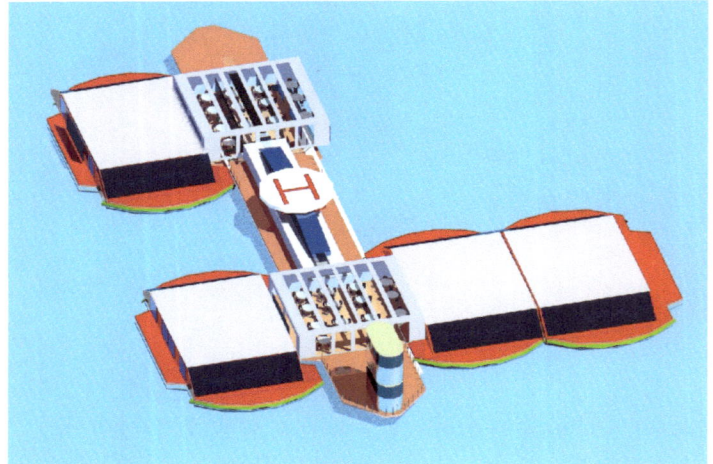

The concept of machine-to-machine transportation— component coupling—is flexible and workable. The engineering is straightforward and the potential benefits are considerable.

The next sub-section outlines the mechanics of the shuttle water taxis that are attached to and move along with the dock-ship—the mechanics isolate the passenger compartment from most wave and wind disturbances acting against the shuttles.

Shuttle suspension as an aid to commuter comfort

Preventing motion sickness on the water and maintaining a stable platform at dock are high priorities in AMS.

Even a short boat ride can be miserable if conditions are rough. This concept provides a means for compensating for water conditions and improving AMS riding qualities.

Water taxis are designed with a passenger compartment riding 'piggyback' on an undercarriage that includes the engines, pontoons and hydraulic and pneumatic suspension.

Water taxis have both hydraulic and pneumatic suspension system components, similar to those components found in Hover-effect ships that substantially isolate the motion of the passenger component from the undercarriage/

Watercraft suspension raises and lowers the passenger cabin structure to fit within the docking structure, as already discussed, but could also function to smooth the ride while the ship was on the water.

The importance of installing multiple types of suspension is that as waves 'oscillate ' the undercarriage the passenger cabin, in contact with the dock or dock-ship superstructure are substantially stationary; passengers have a stable and secure footing while moving to the doors; nor does the water taxi list to one side as the weight of passengers move to one side of the ship.

Much of the operation of a maritime transit system must be automated, because human control is too uncertain, too slow to be effective and safe in fast-moving environmental conditions.

4

AMS-Robotics? Why?

Docking water taxis at night and during inclement weather is work best done through automation. Docking the water taxis would be done water taxi engines under program control so that human operators would not be taxed with the burden of 'parking' a three hundred ton machine into a geometric docking structure at night.

> Automation is used throughout this transportation concept to integrate the overall operation of components and to provide a means for increasing efficiency, reliability, comfort, and safety of passengers and employees.

The basic reason for automation is that some tasks are too complicated or tricky to be done repeatedly and reliably by people. Machines are better suited to certain kinds of work. Any task that is repetitive can be mechanized. (Automated).

Docking (and refueling) of AMS would be a repetitive task. Docking operations and refueling of the ships should be automated.

Automated docking of AMS craft is essential for two reasons. First, human control is less predictable, less reliable, especially in variable operating conditions; and, second, the possibility that hydrogen fuel is used as a primary power source.

Automation in fuel handling is preferable to manual refueling, because automated refueling takes place under computer program control each time an AMS docks. If the docking sequence is already under an automated control it makes sense to further automate the refueling.

When a ship or Automated Maritime Shuttle (AMS) requires refueling, the entire operation takes place below the deck structure, which acts as a fire barrier to all operating equipment above.

Passengers at no time are exposed to open flame from hydrogen in the event of an industrial accident. Passengers are also protected from the unauthorized movement of a trolley or ship while the passenger transfer process was ongoing.

Automation and transportation

Each maritime *Passenger Transfer Facility* has five to eight AMS docked, but only four to six would be in scheduled, routine service. All eight AMS, however, would be warmed up, ready to go.

This provides instant backup transit capability. If an AMS craft is pulled out of service by the control network, a detected fault, and is shut down, another reserve AMS craft is automatically brought 'on line' for instant service.

Automated machines

Service continuation is a necessity. In unconventional transit *some* equipment (bus, trolley car, ship or aircraft) *has* to be available and ready to transfer commuters. Commuters that have arrived *late must be accommodated*. Automation is one way to insure transportation equipment *is always* available.

Modern aircraft have automated navigation systems that result in the pilot, fundamentally, going along for the ride. Even landing is highly automated such that a response from the pilot is necessary only if some event requires the pilot's corrective action.

5

Automation, peace, quiet and low lighting

One of the primary goals of unconventional maritime transportation is to create a 'good neighbor' transit system. Automation helps, because it largely eliminates human error.

The world today has excellent communication capabilities. Much of that communication is between machines, in fact, most of it is between machines. Automated devices that are properly programmed and maintained don't daydream, get angry, or make mistakes.

AMS would have onboard control and communications interfacing with a river traffic control network. The control provides a level of response time and accuracy of response, and a level of consistency not possible with humans.

Constructing transportation that is *quiet and unobtrusive* is actually very important considering that the concept envisions an entire fleet of robotic watercraft (and specialized ships) functioning as connector links between opposite sides of rivers, bays or lakes; or as connector transportation between cities.

AMS- all weather transportation

AMS systems operate under a wide variation in conditions. The ride must be smooth, safe, and the total operation must continue almost no matter what the weather or water conditions. Automation is the key to providing all-weather transit services.

AMS includes a network of environmental sensors

AMS and a control network work cooperatively to program control to monitor river conditions. AMS position and orientation, and docked status are all factored into the calculations. This data is essential to the orderly operation of a large number of AMS operating as a coherent fleet and *not as a crowd of individual ships.*

In-river sensors include low-level radars, and low-power sonar to detect other river traffic. Sailboats, or kids on inter-tubes would not need to fear AMS. The location, speed and direction of other objects on the river would be mapped and that information relayed to the AMS fleet.

Extended season operation

Two factors dominant the design of maritime systems with extended season operation:

- Automation—provides a means for accurate piloting in all weather conditions; and, it provides a means of passenger safety along with greater economy of operation.

- Equipment suitable to the operating conditions.

AMS adjusts to weight-level and dry floors

The passenger cabin structure also elevates up and down, because the *dock structure* is also vertically moving-adjusting to the rise and fall of water elevation.

AMS craft must not only dock securely, but the passenger compartment must elevate to the elevation of the dock so as to maintain a safe, dry, level flooring between the operating components.

Pneumatic mechanisms

Water taxi passenger compartments have pneumatic inflatable members to cushion the walls of the passenger cabin from the structure and weight of the cable car trolley.

In a like manner, a passenger cabin structure skirt, that mates with the docking geometry, also includes a *bumper element* to cushion the movement of the watercraft against the docking structure.

Hydraulic rams raise or lower the passenger cabin to align with the vertical elevation of the dock structure—leveling the flooring between the carrier and the interior of the cable car trolley.

All of the components, the dock structure, water taxis and the trolley elevating mechanisms function *as a unit* to achieve level flooring that is necessary for easy passenger ingress and exit.

Windy and high wave conditions

Water taxis are highly maneuverable with omni-directional jet thrusters to aid in docking. The docking sequence would be automated. Part of that automated procedure would be actuation of sub-surface water jets close to the docking structure.

When wind and wave conditions (or ice buildup) became a threat to navigation or accurately controlling the shuttle, sub-surface air jets would actuate. High-pressure air coming to the surface breaks up the waves causing the water to foam. One such anti-wave device is illustrated on page 59.

Winter Conditions

Ice on most rivers in the lower forty eight states or most of Europe, South America, or Asia is not of sufficient depth to cause problems with ruggedly built ships of the type suggested here: pontoon ships with powerful jet thrusters. Water taxis, without assistance, can break thin sheet ice. Heavier ice, perhaps to two to three inches thick could be broken up with the river shuttle's jet thrusters. Heavier ice, up to six inches thick, can be broken up by hovercraft.

Icy conditions

Automated Marine Shuttles are robotic machines. They operate through the night to keep ice from forming in their individual lanes of travel. Even in a city like Moscow, Russia using powerful water jets to churn the water AMS craft move along the same path and would prevent sheet ice from forming in all except the most extreme conditions.

Fleet operation and conventional river traffic

Water taxis are designed to share the river with all conventional traffic; that is, sailboats and other vessels don't have to worry about 'dodging' robotic ships bearing down on them.

Well, for the most part they won't have to worry about that, because AMS craft would be on a timed cycle. The AMS shuttles (the shuttle fleet) would all leave one side of the river at the same time, and their individual speeds and destinations would be calculated to bring them into other docking facilities at approximately the same time. In other words the shuttles form a fleet; and, the fleet moves out at the same time. They would only be in the river itself for about ten minutes out of any hour.

Infrastructure and Control Networks

All modern transit relies on network control networks for coordinating the movements and speeds of components. Without the regulation afforded by modern telecommunications and control networks modern transportation would not be possible.

A control network integrates AMS fleet operation

A control network does an analytical check of each piece of equipment at each docking sequence. The control network integrates the functioning of the operating components and checks major operating components to insure that the equipment is ready for immediate service.

Part 2

Controlling the aquatic environment

1

Aquatic operations

Operating in marine environments presents special difficulties such as fog, wind, and wave conditions. In actuality controlling the aquatic environment so as to provide a well lighted, secure, ice-free operational zones around dock structures.

The Passenger Transfer Site provides:

❖ Operational support for traffic/river control and

❖ Provides an operational interface to the aquatic control zone so that passengers can move seamlessly from watercraft to other types of transportation such as buses and,

❖ Provide *accommodations* in the event of travel delay.

PTS operational support includes active management of the landing zone so as to insure a safe operational environment.

Wave quelling

Wave quelling devices provide the following operational and safety functions:

❖ Break up wave action so as to reduce turbulence in the landing zone and,

❖ Provide jets of hot water and volumetric air as required to dispense fog near the water surface so as to provide visibility for landing,

❖ Provide compressed air to actuate underwater lighting and safety systems to insure safe aquatic operations and,

❖ Provide signal lighting to departing and arriving ships and watercraft as to a proper routing so as to maintain clearance from other traffic and,

❖ Provide auxiliary electrical and air compression power for crews to work underwater in constructing and maintaining equipment.

A wave-quelling device illustrated on page 59 uses compressed air to break up sub-surface and surface waves so as to create a 'froth' of air and water to smooth waves at the docking structure.

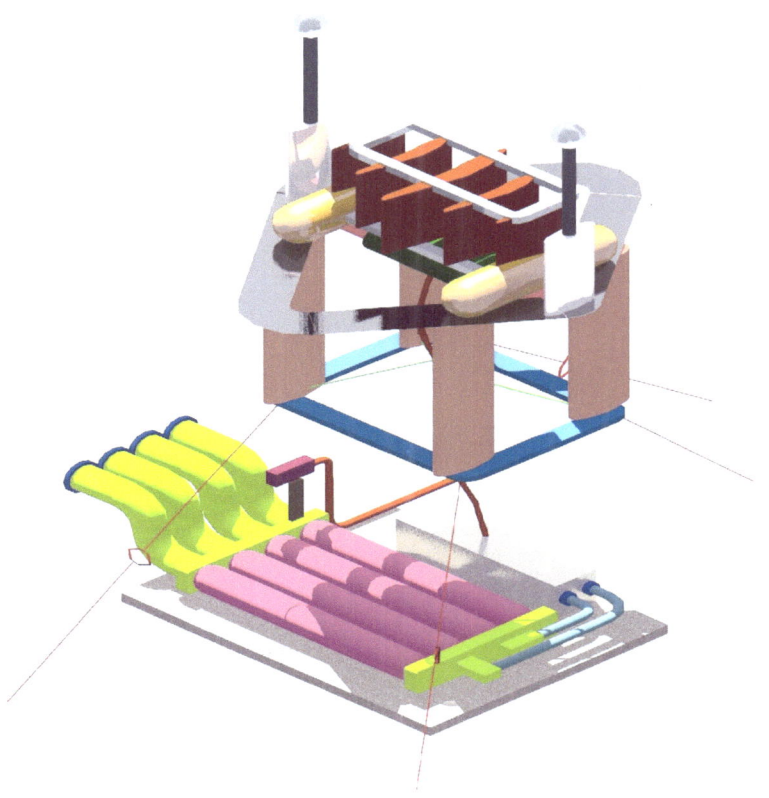

Aquatic operations must be capable of maintaining wind and water conditions within a range: Wind must be controlled to avoid machines operating in close proximity from slamming into, crashing into, each other.

Wind makes waves and the wave height and velocity must also be controlled for same operations. The illustration above is one type of 'heavy duty' machine that pumps air into the water to break up waves and to create a form or froth for safe equipment operation.

2

Multiple guideway systems

The illustration on page 61 shows a maritime *Passenger Transfer Site*. It has a dock structure with berthing bays for water taxis. It also has multiple ship channels.

A Hover-effect ship is berthed in the inner channel with two water taxis docked in the outer channel. The multiple guideways allows two different sets of monorail trolleys to transfer passengers from ships and from the water taxis to operate independently.

On the shore buses wait to accept passengers. The different trolley guideways allows passengers to exit and enter the buses from the right or left—the buses are 'straddled' by two adjacent cable car trolleys.

One maritime Passenger Transfer Site such as this can transfer approximately six thousand passengers each hour of operation.

Here are the numbers: Each trolley carriers 30 passengers and takes five to ten minutes to transfer passengers from a ship to a bus. (That's both ways— sixty passengers)

If the trolley makes 20 trips per hour (it's automated) that means six hundred passengers per hour are transferred. If the trolley is in service for ten hours a day that is 6,000 passengers. Ten trolleys operating during that time equals 60,000 passengers in a ten- hour operation.

Part 3

Energy and maritime transportation

1

Piezoelectric generation of electricity

Innowattech, a research organization in Israel, has developed a new breed of piezoelectric generators. This type of piezoelectric compound salvages energy when deformed or compressive as by a train rolling over the piezoelectric material.

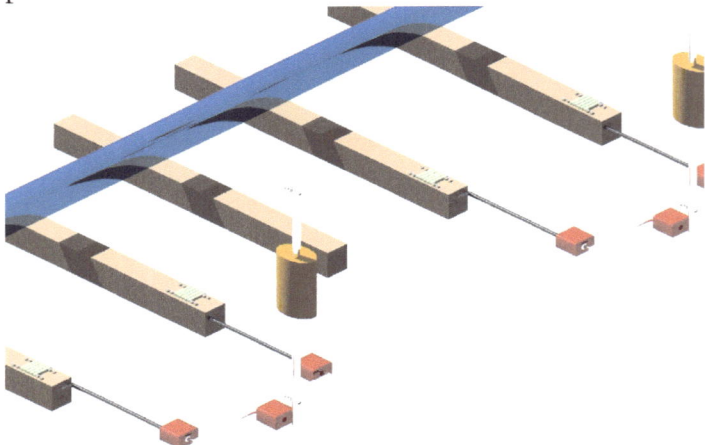

The illustration above shows small (light yellow) components positioned under a railroad rail so that as the train passes over the rail is deformed.

The mechanical energy of deforming the rail is usually lost, but I.P.E.G. piezoelectric compounds salvage much of the otherwise wasted mechanical energy. The illustration on the top of page 63 is a representation of the piezoelectric material.

The Innowattech solution maximizes the 'salvaging' of mechanical energy so as to create a means for generating large amount of useable electricity.

This energy salvaging system is considered to cost less than solar energy and is not dependent on climate or even on temperature. Gravity is reliable. The weight of any device applying sufficient weight or pressure on the piezoelectric compound will salvage mechanical energy that would otherwise be lost.

For example, the illustration on page 64 shows the superstructure of a Hover-effect ship elevated or lifted up from the platform assembly. The Hover-effect ship is designed to be smooth riding due to the separation of the passenger component from the platform assembly. Primarily a plurality of pneumatic cushions in addition to a hydraulic suspension system provides the passengers with a passenger component 'floating' on the platform assembly.

2

Using piezoelectric generators in Hover-effect Ships

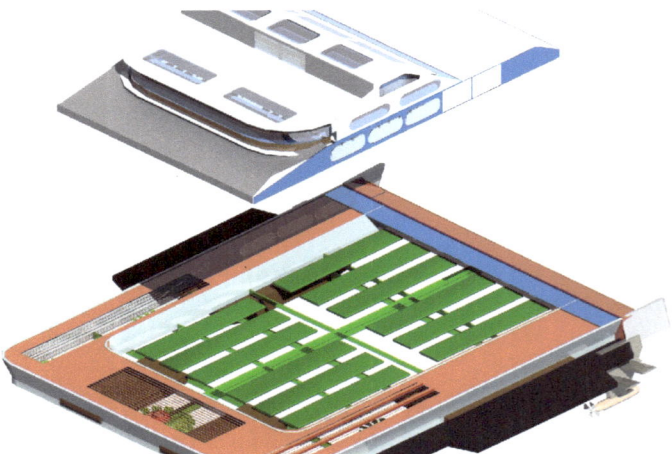

This illustration above elevates the passenger compartment to show the pneumatic and hydraulic suspension system. The illustration below shows a plurality of piezoelectric modules positioned so as to accept the weight of the passenger compartment—the weight of the passenger compartment compresses the I.P.E.G crystals to produce electrical current.

The passenger component might weigh three hundred to four hundred tons so that sufficient electrical energy is produced to meet the internal energy requirement of the ship.

The ability to efficiently salvage mechanical and thermal energy provides us with a useful tool to further the advantages of alternative transportation. The next sub-section looks at using piezoelectric modules with AMS waterbus robotic appliances.

Water taxis as energy co-generators

The Innowattech piezoelectric modules last approximately thirty years. A two-thirds kilometer of railroad is said to produce about 133,00 kW of electricity per hour.

Of course, how much electricity is produced is dependent on the weight or compressive pressure (the factor of deformation) applied to the piezoelectric crystals.

The illustration on page 66 shows Innowattech modules (in yellow) arrayed on the docking element of a water taxi. In this design the water taxi is an automated waterbus appliance that is under automated control when docking.

If the AMS weigh approximately five hundred tons that is about the weight of a metro train, but the AMS has the advantage of geometrically sliding into its docking structure:

1. The moving weight of the AMS has mechanical advantage in that the momentum of the shuttle is 'directed' downward to a smaller area.

2. Hydraulic rams secure the AMS to the dock and also apply additional pressure.

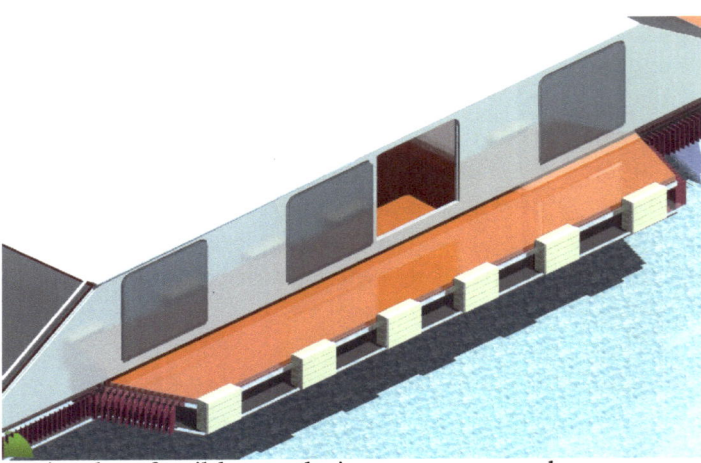

It is also feasible to design a compound system, as shown above, with 'staggered' modules so that as the waterbus docks the weight of the ship pressures against a dock structure deforming piezoelectric modules.

Part of the electric energy produced from the ship's modules is stored in the ship's batteries for internal operations; and, part of the energy produced by the modules on the dock may be used to operate dock computerized equipment and automation.

The illustration on this page shows hydraulic rams, in red, positioned so as to apply mechanical pressure to a water taxis' docking flange.

Hydraulic rams in this configuration are mounted on a carousal element so as to 'capture' the incoming water taxi and 'pivot' with the moving waterbus. The rams, under program control, *guide* the waterbus into a docking slot in the docking structure. This procedure is one of those procedures that would be automated

In the illustration below hydraulic rams guide a water taxi flange into the slot and then applies compressive pressure to secure the ship. The pressure applied on the piezoelectric modules generates an electrical output.

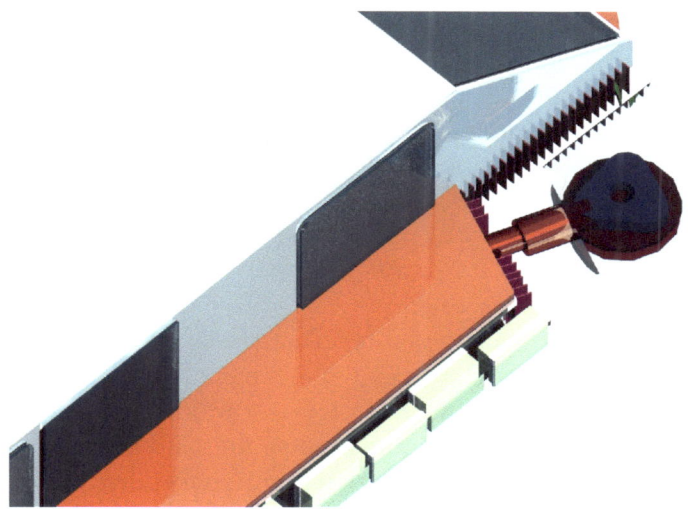

Dock structures located out in Bays or rivers will not need underwater electrical wires strung out to them; they will have a means of generating their own electrical power.

In this system every time a water taxi or ship docks the compressive energy of the watercraft against the dock will generate electricity.

3

Part summary

People that study history are inclined to believe that trains replaced steamboats as the dominant way to travel, because trains are somehow inherently better than ships.

In fact trans replaced steamboats for the simple reason that trains have brakes: A train can stop to pick up a potential passenger waiting by the tracks.

Not so many years ago a person waiting by the tracks could simply 'wave-down' a train. The train would stop, the conducer would collect a fee and the new passenger would take a seat.

Steamboats, of course, could not stop, because if they lost power they drifted onto the rocks. So, the only way to get aboard a steamboat was to travel to the port and get on a ship that was tied up.

Trains won the transportation competition simply because people did not have to travel so far to access that form of transportation.

This work and Unconventional Transportation suggest that a new level of traveling convenience could be achieved by integrating different modalities of transportation into a single functioning unit and that the key to achieving that is to design transportation components that interlock, or connect so that passengers move seamlessly and safely from one mode of travel to another.

There is something else that needs to be said about reorganizing transportation: This concept or something like it would create millions of jobs and reduce our dependency on oil.

A list of references and suggested reading is included for your consideration.

References

Patents

United States patent # 6,497,189, December 24, 2002, issued to Vollmerhausen. Hover-effect Craft. Described in this work as Hover-effect Ship.

United States patent # 6,539,887, April 1, 2003, issued to Vollmerhausen. Bus to Boat Passenger Transfer Facility. Described in this work as a water taxi system.

Additional Reading

Achieving Planned Innovation
Frank R. Bacon, Jr.
Thomas W. Butler, Jr.

The Free Press

The Invisible Web
Uncovering Information Sources Search Engines Can't See
Chris Sherman and Gary Price

CyberAge Books

The Design of Design
Gordon L. Glegg

Cambridge Engineering Series

Beyond Growth
Herman E. Daly
Beacon Press

Other publications by Robert H. Vollmerhausen:

Unconventional Transportation
2nd edition. Completely revised
October 2011

This work describes a bus-train business model for designing alternative transportation.

Railroads, Biomass and Synthetic oil

Using rail rights-of-way to grow biomass and create a sustainable synthetic oil platform.

Novels:

The Sunburst Renegotiation

Shell Game

White Shadow

Earth, Mind and Murder

Jacob Ebbtide

Hidden Jungle

Becky's Flight